Lais Brito Cangussu

Coffee chemistry

Lais Brito Cangussu

Coffee chemistry

Relation to beverage quality

ScienciaScripts

Imprint

Any brand names and product names mentioned in this book are subject to trademark, brand or patent protection and are trademarks or registered trademarks of their respective holders. The use of brand names, product names, common names, trade names, product descriptions etc. even without a particular marking in this work is in no way to be construed to mean that such names may be regarded as unrestricted in respect of trademark and brand protection legislation and could thus be used by anyone.

Cover image: www.ingimage.com

This book is a translation from the original published under ISBN 978-613-9-79711-0.

Publisher:
Sciencia Scripts
is a trademark of
Dodo Books Indian Ocean Ltd. and OmniScriptum S.R.L publishing group

120 High Road, East Finchley, London, N2 9ED, United Kingdom
Str. Armeneasca 28/1, office 1, Chisinau MD-2012, Republic of Moldova, Europe
Printed at: see last page
ISBN: 978-620-6-24994-8

1

Summary

1. Introduction

Coffee is the second most important commercial product in the world, second only to oil. Therefore, it has great economic importance worldwide. Brazil is the largest coffee producer in the world, followed by Vietnam, Colombia, Indonesia, Ethiopia and India. Thus, the economic importance of coffee becomes more expressive in Brazil. According to the International Coffee Organization, in 2017 coffee production in Brazil was 3,090,000 tonnes and in Vietnam it was 1,530,000 tonnes. With these data, it is possible to observe that coffee production in Brazil was practically double that of the second largest producer. We know that the countries that consume the most coffee are not the largest producers, with the exception of Brazil. With this reality, Brazil has a giant international market to serve. As Brazil is the largest producer and the largest consumers do not produce or have irrelevant production, any international crisis affects the price and sale of coffees in Brazil.

International markets are more demanding in relation to coffee quality, so it is essential that Brazilians have more technical knowledge about this product. Through knowledge it is possible to have high quality coffees and, consequently, reach more demanding markets. Thousands of small producers depend on coffee for their livelihood, which reinforces the importance of technical knowledge about this product that is so important for our country's economy. The state of Minas Gerais predominates coffee cultivation in Brazil, showing greater relevance of coffee in this state.

Currently, national consumers are undergoing a change in their coffee consumption behaviour, reflecting the increase in demand for gourmet and specialty coffees. These coffees are produced through a greater selection and monitoring of the coffees.

beans, roasting and grinding process. The first step to understand the transformations that occur in the coffee bean during roasting is to know how it is formed. Coffee comes from the *Coffea* plant *(*Figures 1, 2 and *3),* which is a shrub with bright green leaves that produce white flowers, which develop into the fruit. The fruits (Figure 4) are composed of an outer skin called the rind, which is red or yellow in colour. It is followed by the pulp fraction, a yellowish, fibrous bed. After the pulp, there is the mucilage, a translucent layer rich in carbohydrates. It then contains the parchment, a fraction that lies loosely between the mucilage and the silver film. The silver film is attached to the seeds and will often only come off at the time of roasting due to the expansion of the grain. In the middle are the two (mostly) central seeds, only these are used to produce the beverage we call coffee.

Figure 1. Source: Murthy & Naidu, 2012.

Figure 2 Source: Murthy & Naidu, 2012.

Figure 3. Source: Murthy & Naidu, 2012.

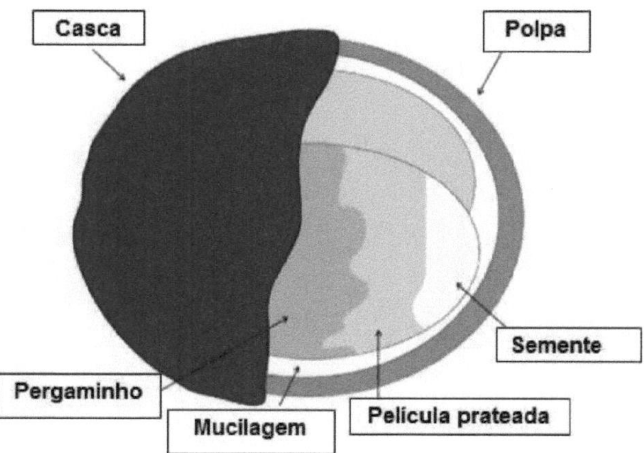

Figure 4 Source: Esquivei & Jiménez, 2012.

If only the two centre seeds are used for the production of the beverage, all other fractions become waste. These residues are rich in bioactive compounds (compounds that bring health benefits), such as trigonelline, caffeine and phenolic compounds, which can be used by the food, pharmaceutical and cosmetic industries to make various products. However, currently, these residues have no economic value and are often not disposed of in the correct way, causing environmental damage. Thus, it is important that more studies are done with these residues, being a way to add value to this material and reduce the environmental impact they cause.

There are two species of coffee with economic value worldwide: Arabica coffee and Robusta coffee. Arabica coffee is known for its superior quality. This species has higher acidity. Robusta coffee has a higher amount of caffeine and more bitterness. Arabica coffee is more expensive and more susceptible to pests. Due to the high price of Arabica coffee, industries make *blends* with the two species, which consists of mixing them, obtaining a final product with the combination of the characteristics of both. However, the best coffees are made with

100% Arabica beans, since the robusta species impairs the final quality of the drink.

The next chapters will discuss the compounds present in coffee and the changes they undergo during roasting, forming the quality attributes of the drink.

2. Water and Minerals

Water is formed by the covalent bond between one oxygen and 2 hydrogens. It is a fundamental component of food, being present in high quantities in fruits and vegetables. Some physicochemical properties of water are shown in Table 1.

Table 1 - Physico-chemical properties of water

Melting point (change from solid to liquid)	**0 °C**
Boiling point (change from liquid to gaseous state)	100 °C
Density	$1 g/cm^3$
pH	7,0

The melting and boiling point of water is high compared to other compounds and what justifies this fact is the ability of water to form hydrogen bridges, which are difficult to break, resulting in greater difficulty in passing from one state to another. Hydrogen bridging occurs with water molecules because oxygen is very electronegative, meaning it brings the shared electrons closer to itself (creating negative charge). One water molecule can form hydrogen bonds with up to four other water molecules. Because of this, water has surface tension, which allows a mosquito to walk on water,

Coffee cherries (fruit) contain between 50-75% water. As water plays a role in product deterioration, mould growth, mycotoxin production, fermentation and physical, chemical and sensory changes, it is necessary that its quantity is reduced in coffee. In this way, coffee is dried until it reaches between 10- 12% humidity. The amount of water allowed in coffee varies from country to country. In Brazil, the legislation allows raw coffee to have a maximum of 12.5% moisture and roasted coffee a maximum of 5%. When drying, the intention is to eliminate as much free water as possible from the food. Free water is water that is available for use by microorganisms, enzymatic reactions and is used as a solvent. In addition to free water, there is bound water, which has impaired mobility. This water is tightly bound to solutes and macromolecules and is not available for microorganisms to act and for reactions to take place. Water activity is used to measure the availability of water in a food. Pure water without nutrients has a water activity of 1.0 and the more the amount of nutrients in the water increases, the more the water activity decreases.

To determine the amount of water in coffee, either the direct method or the indirect method can be used. The direct method measures the actual water present in the product through its evaporation by heat. The sample is placed in an oven with air circulation (important so that the air does not saturate) and left to dry for a certain period and temperature (example: 24 h at 132°C). Before placing the sample in the oven, it is weighed and after drying, it is weighed again. The difference in weight obtained is the amount of water evaporated. In the indirect method, electronic meters are used, which do not directly measure the amount of water present but rather some property that water presents, such as electrical capacitance. The greater the amount of water, the greater the flow of electric current.

Minerals are inorganic materials, i.e. they do not contain carbon, hydrogen, oxygen and nitrogen. They are present in low quantities in food, however, they perform essential functions for living beings and food. The main minerals that are essential for human health are calcium, potassium, magnesium, sodium, phosphorus, chlorine, iron and manganese. About 1 gram per day of the essential minerals should be consumed, and excess can lead to toxicity. Excess minerals are rarely present when a varied diet is consumed. An important factor to consider is the bioavailability of minerals, which means the amount of the mineral consumed that is absorbed by the body. For example, the bioavailability of iron is about 1% and that of sodium and potassium is about 90%, so iron deficiency is easier to acquire than sodium and potassium.

The amount of minerals (RMF - fixed mineral residue) present in coffee varies according to origin, soil fertility, genetic influence and processing. In the coffee hulling stage, minerals are lost, as they are concentrated in the outer layers that are removed, leaving only the seeds, which are poor in minerals. Raw coffee can have a maximum of 5% RMF and roasted coffee 3%. The mineral present in the greatest quantity in coffee is potassium (about 40%). It is essential for humans, being important for muscle contraction, heart rate and nerve conduction. However, potassium deficiency is rare, since it is consumed more than necessary and its bioavailability is high. To a lesser extent, calcium, magnesium, copper and manganese are also present in coffee, with copper having the property of stabilising foam, which is interesting for coffee.

The amount of minerals in a food is determined by ash, which is obtained after combustion of organic compounds. It is possible to determine minerals by this method because they are indestructible, not being destroyed by heat, light, extreme pH and oxidisers. Therefore, by placing the food in an oven (muffle furnace) at a high temperature (550°C) all organic matter is destroyed, leaving only the minerals (ash).

3.Carbohydrates

Carbohydrates account for more than 90% of the dry matter of plants, i.e. after removal of water it is the component present in the greatest quantity in plants. They are divided into monosaccharides, oligosaccharides and polysaccharides. Monosaccharides are free sugars with low molecular mass and when they come together they form oligosaccharides and polysaccharides. Polysaccharides are the most complex carbohydrates and are found in greater quantities in foods.

Carbohydrates are important for the development of coffee flavour and aroma and in the formation of pigments and other high molecular weight products. In coffee, they consist of polysaccharides (40-50%), low molecular weight sugars and carbohydrate derivatives such as pectin and lignin. Among the low molecular weight sugars present in coffee, sucrose is the main one. Sucrose is a disaccharide (composed of two monosaccharides), formed by one molecule of glucose and one of fructose. It is table sugar, which we use daily to sweeten recipes. Our table sugar is produced by sugar cane, but in other countries it is produced by sugar beet.

Arabica coffee (~8%) has twice as much sucrose as robusta coffee (~4%), most of the time. Carbohydrates are important in the formation of pigments, flavour and aroma of coffee because they participate in a reaction that occurs during roasting, called the Maillard reaction. This reaction occurs whenever there is the presence of reducing carbohydrate and protein or amino acid at high temperature. Exactly what happens with coffee beans when subjected to the roasting process. During this reaction that occurs between the reducing sugar and the protein/or amino acid, compounds responsible for the aroma, flavour and darkening of the bean are formed. The browning formed by this reaction is classified as non-enzymatic browning, since the reaction occurs without the action of enzymes. Despite the desirable sensory characteristics formed by this reaction, it also forms a carcinogenic compound called acrylamide. Coffee is not the food that has the highest amount of acrylamide. French fries, for example, have up to 6 times more acrylamide than coffee. One explanation for coffee not having a very high amount of acrylamide is that it is destroyed at very high temperatures (above 220°C) and in some roasts this temperature is reached. There are some techniques capable of reducing the amount of final acrylamide in the product: increasing pH and bleaching. By increasing the pH, the degradation of this substance is favoured. Blanching vegetables reduces the amount of acrylamide formed, since amino acids are solubilised in the blanching water and as the Maillard reaction only occurs in the presence of proteins and amino acids, the fewer amino acids, the less acrylamide is formed. Blanching is a pre-treatment, where the vegetable is placed for a few seconds or minutes in hot water and then rapidly cooled. This method inactivates enzymes, which are responsible for browning reactions.Another reaction that occurs with carbohydrates at high temperature is caramelisation. In this case there is no need for proteins and/or amino acids. Caramelisation is easily observed when table sugar is put on the fire and turns into caramel. As it is noticed in sugar, darkening occurs during this reaction. The same occurs with coffee during roasting. Caramelisation causes browning and the formation of flavour and aroma compounds. This reaction is facilitated in the presence of acids and salts.We have seen that Arabica coffee has a higher amount of sucrose, but it has a lower amount of total reducing sugars (~0.1%) compared to Robusta (~0.4%). The reducing sugars present in coffee are glucose and fructose and traces of stachyose, raffinose, arabinose, mannose, galactose, xylose, ribose and rhamnose. During roasting of the coffee bean, sucrose is lost rapidly, whereby in a lightly roasted coffee, only 3-4% of its original sucrose content is retained. In dark roast, sucrose is lost completely. Reducing sugars will be formed during roasting by hydrolysis of polysaccharides or oligosaccharides, since reducing sugars form the more complex sugars and when they are broken down, simple sugars are released. Although reducing sugars are formed during roasting, their quantity is not greater in roasted coffee, as they are also degraded by the heat of roasting. Among the reducing sugars, glucose is less susceptible to loss on roasting and arabinose is more susceptible. Polysaccharides are more resistant to the roasting process, with on average 75% of their initial composition remaining in the roasted coffee. It cannot be said which coffee species has the highest amount of polysaccharides, since the difference of these compounds between them is minimal.The polysaccharides present in greater quantity in coffee are: mannans, galactomannans, arabinogalactans, cellulose and xyloglucans. Polysaccharides have an influence on the body attribute of the beverage, since they are the main responsible for the viscosity of coffee extracts. However, home-brewed coffee does not extract polysaccharides as much as espresso coffees do, as they are extracted with high pressure. They are also responsible for the stability of the espresso foam and the retention of volatile compounds. The volatile compounds that form the aroma compounds of coffee, so polysaccharides are important in preserving the aroma of coffee, since they bind to the volatile compounds. The determination of carbohydrates can be done using a refractometer, obtaining the °Brix measurement. The °Brix is used to measure the amount of soluble solids present in the solution, more specifically the amount of sugars. Another method is using a spectrophotometer, which measures the absorbance value of the sample in relation to a radiation source at a specific wavelength. In order to use this method, a calibration curve with a carbohydrate standard must be performed. This method will quantify the total carbohydrates in the sample. If the interest is to know which carbohydrates are present, it is necessary to use other techniques. The gas chromatography technique is widely used for the determination of monosaccharides. In this case, it is first necessary to hydrolyse the glycosidic bonds, i.e. to break the bonds between the monosaccharides that form the polysaccharides, in this way the monosaccharides are released. Gas chromatography can only read volatile compounds, so it is necessary for the monosaccharides to be transformed into volatile compounds in order to read them. The volatile compounds formed by monosaccharides are alditol acetates, with each monosaccharide forming a specific alditol acetate. After the formation of the alditol acetates, they are read in the gas chromatograph (Figure 5) and then it is possible to obtain the percentage of each monosaccharide separated. But it is necessary to read the individual standards of each monosaccharide to identify the retention times (Figure 6). With the results obtained it is possible to deduce which polysaccharides are present in the samples, but to be sure it is necessary to combine the results with other techniques such as infrared, nuclear magnetic resonance and mass spectrometry.

Figure 5

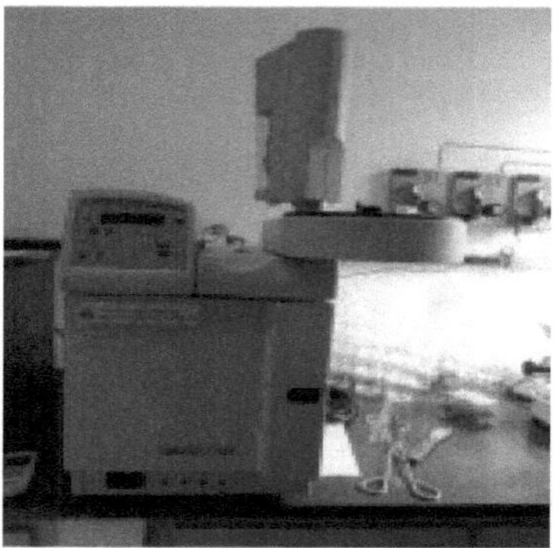

Figure 6 - 1-Erythritol, 2- Deoxyribose, 3-Ramnose, 4-Fucose, 5-Ribose, 6-Arabinose, 7- Xylose,

8- Deoxyglucose, 9- Allose, 10- Mannose, 11- Galactose, 12- Glucose, 13- Myo-Inositol.

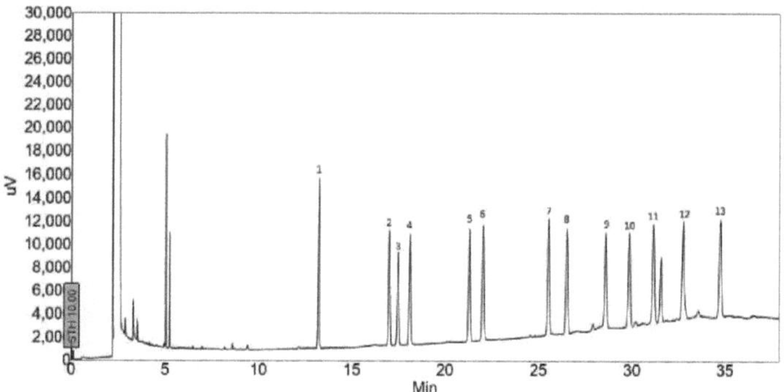

4.Nitrogen compounds

Nitrogenous compounds are compounds that contain nitrogen in their composition. The most important nitrogenous compounds in coffee are proteins and amino acids, trigonelline and nicotinic acid and caffeine. Amino acids form proteins. They are composed of carbon attached to an amine group, a carboxyl group, a hydrogen and a radical. One amino acid differs from another only by the nature of the radical. Amino acids are joined by peptide bonds to form proteins. The peptide bond is formed between the amine group of one amino acid and the carboxyl group of another amino acid, releasing a water molecule.

Proteins can have one or more structures: primary structure, secondary structure, tertiary structure and quaternary structure. The primary structure is linked by covalent bonds (electron sharing) and the secondary structure by hydrogen bridges (hydrogen atom covalently bonded with an electronegative atom - N, O or S). Electronegativity is the ability of the element to bring shared electrons close to itself, thus forming negative and positive poles, where one repels the other. The tertiary structure is formed by hydrophobic bonds, due to the presence of hydrophobic and hydrophilic groups. The hydrophobic groups are positioned in the centre of the protein (avoiding water) and the hydrophilic groups are positioned in the outer parts binding to water. Another bond present in protein structures is the disulfide bridges, a bond that occurs between the sulfur atoms of cysteine amino acids.

Proteins are denatured (loss of their biological activity and functionality) in the presence of some factors: temperature, pH, organic solvents, hydrostatic pressure, among others. During roasting, coffee proteins are denatured by temperature, thus, coffee loses all its enzymatic activity. Enzymes are proteins that are essential for certain reactions to occur.

In coffee, the protein fraction is divided into a water-soluble fraction (albumin) (~50%) and a water-insoluble fraction (~50%). There is no difference in the amount of protein in arabica and robusta coffee, but robusta coffee has a higher amount of amino acids. Asparagine is the main amino acid in coffee beans, being the main precursor of acrylamine (formed in the Maillard reaction). The accumulation of asparagine occurs mainly during normal physiological processes, such as seed germination, and is therefore difficult to avoid. Amino acids, as well as enzymes, are totally lost in coffee roasting. Protein loss, on the other hand, will depend on the degree of roasting, and in severe conditions, the loss can be greater than 50%. Although protein losses occur during roasting, when protein quantification is analysed by the official method (Kjeldahl), there is no difference in composition. This fact occurs because this method detects the amount of nitrogen present in the sample and not the amount of protein and there are other compounds that contain nitrogen in their composition in coffee. During the roasting of coffee, several volatile compounds are formed that contain nitrogen in their composition, so they also enter the quantification of proteins by the Kjeldahl method. There are other methods for protein determination, but the Kjeldahl method is considered the official one.

Proteins and amino acids contribute to the colour and aroma of roasted coffee due to their participation in the Maillard reaction. Proteins are also responsible for increasing the body of coffee, since they are high molecular weight compounds, making the drink more full-bodied. In espresso coffees, greater extraction of high molecular weight compounds is achieved, contributing to a coffee with more body.

Caffeine (Figure 7) is also part of the nitrogenous compounds present in coffee. It is an alkaloid with a high antioxidant capacity, having a greater effect than ascorbic acid. Alkaloids are substances of plant origin that have a basic character and bitter taste, being present with the function of defence against insects and predatory animals. Often, the main reason for drinking coffee is the stimulating effect that caffeine has, but many are unaware of the numerous health benefits it can bring. Other beneficial effects of caffeine are: reducing the risk of diabetes, hypertension, risk of cardiovascular and neurodegenerative diseases, improving mood, reducing symptoms associated with Parkinson's disease, protection against UV rays, analgesic effect, increasing blood circulation, stimulating hair growth, among others. Due to these effects, caffeine is widely used in drug and cosmetic formulations. However, despite the immense health benefits of caffeine, studies also show that it is toxic to many vital processes. However, what will determine whether caffeine will benefit or harm health is the amount ingested. An intake of 75-300 mg of caffeine per day is sufficient to achieve health benefits, corresponding to the consumption of *A to 3 cups of coffee per day. The presence of 50 mg/mL of caffeine in the blood can already be fatal. However, to achieve this amount with coffee intake is difficult, since in 1 litre of coffee there is approximately 1000 mg of caffeine and coffee is the food that contributes most to caffeine intake.

Figure 7. Source: Sigma-Aldrich

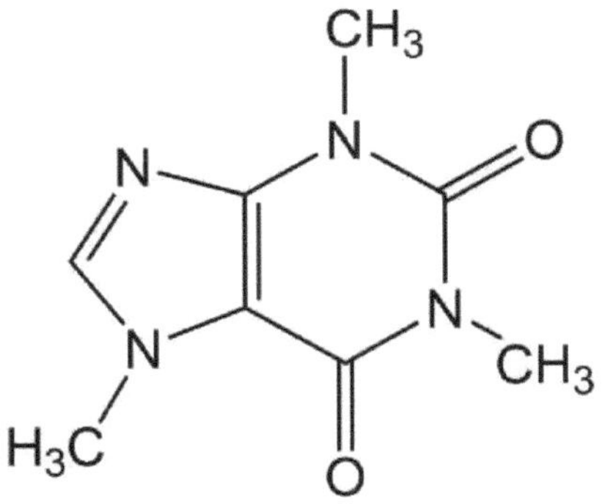

Caffeine is present in greater quantities in robusta coffee, one of the reasons why this species has a more expressive bitterness than arabica. Caffeine is odourless and has a very characteristic bitter taste. Brazilian legislation determines that roasted coffee and soluble coffee must have at least 0.7% caffeine, decaffeinated coffee at most 0.1% and green (raw) coffee at least 1%. Coffees with lower amounts of caffeine may have been adulterated with the addition of other cereals or other substances. Caffeine shows significant losses only in dark roasts, as it is stable in light and medium roasts.

Trigonelline, unlike caffeine, is widely distributed in the plant kingdom. It also has health benefits such as anti-cancer action, memory stimulant, production of vitamin B3, action against tooth decay and antibacterial action. Its toxicity is very low compared to caffeine. Trigonellia makes coffee the only product that in drastic situations such as roasting produces an essential vitamin for humans, niacin (vitamin B3). Green coffee has a low amount of niacin, showing no significant physiological effect. After roasting, coffee starts to present the physiological effects of vitamin B3. The formation of niacin occurs due to the demethylation of trigonelline (Figure 8), that is, the loss of the methyl group of trigonelline. But not all trigonelline is degraded into niacin. The proportion of trigonelline that degrades to niacin is small, and most of the other products are volatile compounds (responsible for flavour).

Figure 8. Source Sigma Aldrich

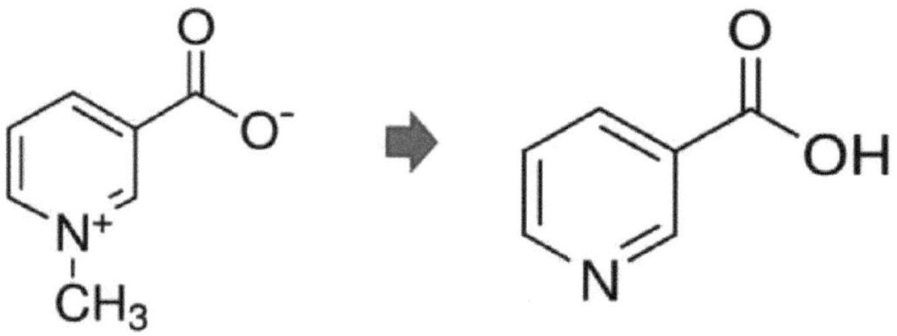

Unlike caffeine, the Arabica species has a higher amount of trigonelline than the Robusta species. Thus, the arabica species will also have a higher amount of vitamin B3, as it is formed through trigonelline. There does not seem to be any correlation between the perceived quality of coffee and the content of nicotinic acid (vitamin B3). Nicotinic acid is very stable in neutral, acidic and alkaline media, presence of oxygen, light and heat, so it is not lost during roasting. Among the benefits of niacin are: cholesterol control, skin health, energy production, arthritis improvement, cancer prevention and diabetes improvement. Lack of niacin causes poor digestion, canker sores, tiredness, depression, vomiting and pellagra.

Both caffeine and trigonelline are easily extracted in the beverage as they are very soluble in water, trigonelline being even more soluble than caffeine. Trigonelline is highly sensitive to the roasting process, and the more drastic the roasting (binomial time and temperature), the lower the amount of trigonelline. During roasting, trigonelline loses 50-80% of its initial amount. Trigonelline has little direct influence on the quality of the coffee drink, as it has a weak bitter taste. However, it contributes to the aroma through the formation of degradation products during roasting, such as furans, pyrazines, alkyl-pyridines and pyrroles. Determination of caffeine and trigonelline can be done using high performance liquid chromatography technique. This technique is similar to gas chromatography, but the mobile phase in this case is liquid and no volatile compounds are detected. Figures 9 and 10 contain chromatograms and spectra of trigonelline and caffeine.

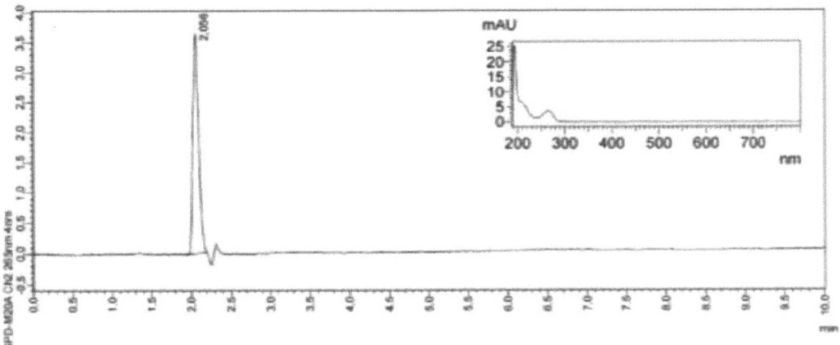

Figure 9 - Chromatogram and spectrum of trigonelline

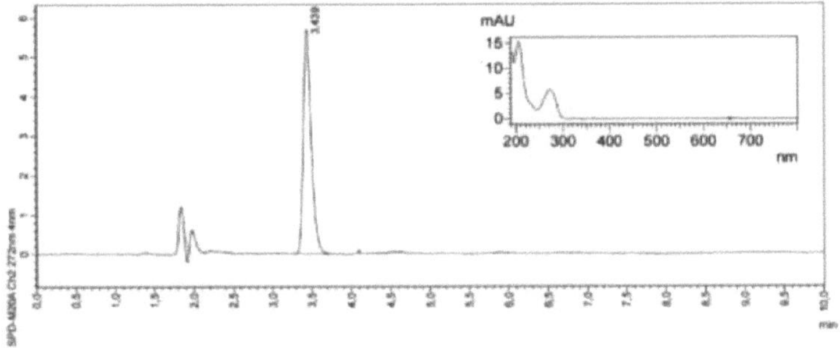

Figure 10 - Chromatogram and spectrum of caffeine

5.Lipids

Lipids are compounds that are soluble in organic solvents and insoluble in water. They can be divided into fats and oils. Oils are mostly liquid at room temperature and fats are solid. Fatty acids are the main components of lipids. They contain an open (aliphatic) chain and a carboxylic group (COOH). Fatty acids can be saturated or unsaturated, depending on the presence or absence of double bonds. Saturated fatty acids have no double bonds and unsaturated fatty acids have double bonds. If they have only one double bond, they are called monounsaturated and if they have more than one, they are called polyunsaturated. Saturated fats are solid at room temperature and unsaturated fats are liquid. Due to the controversy over the consumption of saturated fats, trans fats were created, which are modified unsaturated fats. These fats undergo a process called hydrogenation and change from their cis to trans form, from being liquid to being solid at room temperature.

Free fatty acids are not common in living tissues as they are cytotoxic (damage cell membranes). When they are esterified with glycerol, their toxicity decreases. Fatty acids bound to glycerol form triglycerides. Another type of fat are sterols, when of animal origin they are called cholesterol and when of plant origin phytosterols. Phytosterols are beneficial to health because they compete with cholesterol in the gut, so the body absorbs it instead of cholesterol. Another lipid found in vegetables is waxes. They are found on the surface of plant and animal tissues and have the function of inhibiting water loss or repelling water. Waxes are added to fruits to slow down their dehydration in storage.

Lipids are relatively poor conductors of heat and their melting point increases with increasing chain length, being higher for saturated fatty acids and trans unsaturated forms than for cis. The more double bonds present, the lower the melting point. Oils have a density between 910-930 kg/m^3 at room temperature and fats between 1000-1060 kg/m^3 .

The lipids of green coffee are composed of an oil fraction present in the endosperm, functioning as an energy reserve, and a small amount of wax located in the outer layers of the bean. Triacylglycerols make up the main class of lipids in coffee, while free fatty acids are present in small amounts, corresponding to about 1% of total lipids. Green coffee oil is used in the cosmetics industry thanks to its emollient property (moisturises the skin) provided by fatty acids and diterpenes, ability to block sunlight harmful to human skin health and good anti-oxidant activity. Robusta coffee (9-13%) has a lower amount of lipids than Arabica coffee (14-16%). In roasted coffee there is an increase in lipids, because in addition to oils being released, there is a centesimal increase due to the degradation of other components (net loss of materials in the form of CO_2, water vapour and volatile compounds during the degradation of polysaccharides, amino acids and chlorogenic acids), thus decreasing the percentage of other compounds and consequently increasing lipids, as they are not lost in roasting.

Lipids can contribute to the loss of sensory quality of roasted coffee during storage due to the hydrolysis of triglycerides (lipid degradation), with the release of free fatty acids, which are oxidised. Lipids can undergo hydrolytic and oxidative rancidity. Hydrolytic rancidity occurs in the presence of moisture and oxidative rancidity in the presence of atmospheric oxygen. Atmosphere and temperature influence these reactions when associated with storage time. The use of inert atmosphere and low temperature contribute to a slower loss of free fatty acids, which will be responsible for unpleasant taste and aroma due to the oxidation they undergo.

Roasted coffee oil is a viscous, brown liquid characterised by an intense and pleasant coffee aroma. The colour of the product is mainly due to the presence of fat-soluble products of the Maillard reaction. They are largely used as a source of flavouring for food and cosmetics. Of the fat-soluble vitamins, coffee has small amounts only of vitamin E and cannot be considered as a source of this vitamin. However, the amount of vitamin E present in coffee can be used to identify coffees adulterated by the addition of other cereals, such as corn and rice.

In most coffee drinks, the lipid content is relatively small, but in espresso coffee the content is higher. This is because coffee is prepared with water and lipids are not soluble in water. Hot water can extract a small amount of lipids, while espresso coffee extracts more of them due to the increased pressure. Lipids are carriers of aroma and possible foam destabilisers. Coffee wax is responsible for the unpleasant sensation after drinking and it is desirable to remove it from the bean. It can be used as a natural antioxidant in food and as a source of serotonin. Removing the wax from coffee considerably improves the quality of the drink.

The determination of lipids can be done by hot or cold extraction. In hot extraction, the most widely used method is the Soxhlet method. It is based on semi-continuous extraction, using solvent that accumulates in the extraction chamber for 510 minutes, circulates through the sample and via siphons returns to the boiling flask. The fat content is measured by mass loss of the sample or by the mass of fat removed. The disadvantages of this method are that it is only used for solid samples, it uses a large amount of solvent, the solvent can saturate (not extracting all the lipid from the sample) and the extraction time is long. An example of a cold method is the Bligh & Dyer method, in which the wet sample is homogenised with a mixture of chloroform and methanol in such a proportion that it forms a single-phase system. Then, a dilution with chloroform and water promotes the separation into two phases. The chloroform phase contains all the lipids.

6. Carboxylic acids

Organic acids are important for respiratory metabolism and as reserve compounds in fruits and vegetables. The accumulation of organic acids leads to an increase in sour or acidic flavour. Acidity can be determined by the hydrogen ion concentration (pH), which is related to the degree of ionisation or dissociation of a particular acid present in an aqueous solution of the acid or mixture of acids. Changes in pH can therefore lead to changes in flavour character as well as acidity.

The ideal pH of the coffee drink is between 4.95 and 5.20. The intensity of coffee acidity is related to factors such as fermentation levels, climatic conditions during harvesting and drying, ripening stages, place of origin and form of preparation. The fermentation of coffee beans contributes to the increase in acidity because during its process the production of alcohols and acids occurs, through varied biochemical routes.

The types of acids found in coffee infusions are aliphatic carboxylic acids and some carbocyclic acids. Carboxylic acids are those that have a carboxyl group (COOH) in their composition. The presence of inorganic acids such as phosphoric acid is also reported. The acid fraction of coffee has non-volatile acids such as oxalic, malic, citric, tartaric and pyruvic and volatile acids such as acetic, propionic, valeric and butyric. Depending on the concentration of these acids present, they can impart a pleasant or unpleasant odour and taste to the beverage. Although it is not yet clear which compounds are responsible for the perceived acidity, it is known that citric, malic, acetic, quinic and chlorogenic acids are the acids in greater quantity in the coffee bean, and may be responsible for the acidity of the beverage.

sensory.

Citric and malic acids are the most widely occurring and abundant acids in vegetables. The acidity of fruits usually decreases with ripening due to the utilisation of organic acids in respiration or their conversion to sugars. Table 2 shows the predominant acids in fruits.

Table 2: Predominant acids in fruits.

Predominant acid	Product
Citrus	Lemon, orange, gooseberry, fig, guava, pineapple, pomegranate, raspberry and strawberry.
Isocitric	Mulberry
Malic	Apple, apricot, banana, cherry, grape, peach, pear and plum
Oxalic	Spinach
Quinic	Kiwi
Tartaric	Avocado and grapes

Carboxylic acids are present in smaller quantities than chlorogenic acids in green coffee. The main ones found in coffee are: citric, malic, oxalic and tartaric. The acidity perceived in coffee has always been recognised as an important attribute of its quality. Citric and malic acids are related to the desirable acidity of coffee, with Arabica coffee having higher concentrations of these acids. Improper or undesirable acidity may come from excessive fermentation of the fruit. Thus, raw coffee stored for a long period of time is slightly more acidic than freshly harvested beans.

During coffee roasting, there is an increase in acids, since, similar to other oilseeds, acid is released by enzymatic hydrolysis of lipids. Another factor contributing to the increase of acids in coffee at the beginning of roasting is the reduction of carbohydrates into carboxylic acids and carbon dioxide. Thus, there is an increase in titratable acidity after roasting coffee beans, indicating the formation of acids. The highest concentration of carboxylic acids in roasted coffee is observed at a roasting with a weight loss between 14 and 16%. However, with increasing roasting, the acids are degraded. Thus, the higher the degree of roasting, the less acidic the coffee will be, and the roasting point of coffee beans can be indicated by the final content of acids present. Around 34 aliphatic acids have been identified in roasted coffee (15 volatile acids and the rest non-volatile). Arabica coffee has higher acidity than conilon coffee.

Acidity can be determined by pH measurement (using pH meter or strips) or by titratable acidity determination. Titratable acidity is measured using a pH indicator (phenolphthalein), which is placed in the sample. The pH indicator changes colour according to the pH of the medium. The sample containing the pH indicator is titrated with basic solution until the pH of the solution changes. When changing from acidic to basic pH, the sample will change colour, which indicates the end of the titration. The acidity is determined by the amount of basic solution used in the titration, and the more of this solution used, the more acidic the sample. The pH measuring strips are composed of pH indicators, each colour indicating the pH of the solution.

7. Phenolic Compounds

Phenolic compounds comprise a diverse group of bioactive compounds found in nature. They are widely distributed as secondary metabolites in plants, and are therefore part of the human diet. Currently, about 10,000 phenolic compounds have been described in the literature and, because they are highly antioxidant, there has been great interest in studying them.

Phenolics have at least one aromatic ring in their structure and hydroxyls attached to the ring structure as substituents. Figure 11 shows an example of a phenolic compound (chlorogenic acid). Phenolic compounds are able to slow down ageing and the onset of diseases, and even prevent them. It is known that, in addition to performing protective functions due to their antioxidant properties, they also contribute to the sensory qualities of vegetables such as colour and astringency. Phenolics are divided into tannins, phenolic acids and flavonoids. Flavonoids, in general, have yellowish pigments. Phenolic acids are the simplest phenolics found and tannins are the most complex. Tannins are responsible for the astringency associated with phenolic compounds.

Figure 11 Source: Sigma-Aldrich

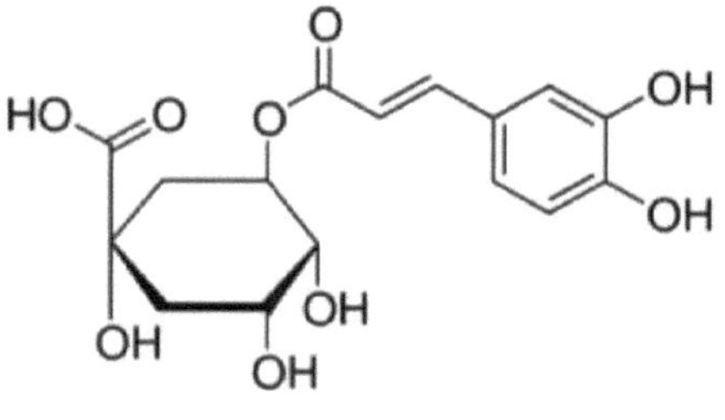

The presence of phenolic compounds in coffee directly affects the organoleptic characteristics of the beverage, such as changes in colour, flavour and aroma during the roasting process. The degradation of phenolic compounds causes the formation of pigments and aroma and flavour compounds. Among the phenolics present in coffee, chlorogenic acids are considered to be of greater relevance and are present in greater quantity in the bean. They are a set of five main groups of phenolic compounds and their isomers formed mainly by the esterification of quinic acid with one of the following acids: caffeic acid, ferulic acid or p-coumaric acid. These groups are: caffeoylquinic acids, dicaffeoylquinic acids, feruloylquinic acids, *p-coumaroylquinic acids* and caffeoylferuloylquinic acids.

During roasting, coffee may present lower acidity due to the destruction of chlorogenic acids that are attached to the bean matrix. Chlorogenic acid is degraded releasing caffeic acid, quinic acid and others. Thus, chlorogenic acids decrease in roasted coffee, and the less roasted the coffee, the higher the acidity. Caffeic and quinic acids are partially hydrolysed, isomerised or degraded to low molecular weight compounds that contribute to flavour and aroma. It is evident that the quinic acid content increases continuously as chlorogenic acid decreases during roasting.Robusta coffee (3.9 - 4.6%) contains a higher amount of chlorogenic acids than Arabica coffee (1.2-2.3%). Chlorogenic acids, as phenolic compounds, have an important role for human health. Foods with significant amounts of chlorogenic acids are highly targeted as they have several proven biological functions. Studies have revealed that they have several pharmacological properties such as ability to increase hepatic glucose utilisation, antioxidant activity, antispasmodic activity, inhibition of HIV-1 integrase and inhibition of mutagenicity of carcinogenic compounds.

8. Volatile Compounds

Volatile compounds are the compounds responsible for the emission of coffee odour. The contact of these compounds with the receptors located in the olfactory epithelium in the upper parts of the nasal cavity form the volatility. The quality of coffee is assessed largely on the basis of its aroma and *flavour* by expert coffee tasters. Most volatile products are derived from non-volatile components of the raw bean, which break down and react during roasting to form a complex mixture. Green beans are considered non-aromatic, but contain the large number of chemical precursors of these compounds, such as sucrose and other carbohydrates, chlorogenic acids, proteins, etc.

The aroma of coffee is important for differentiating species and for the commercial value of the product. Some factors can influence the sensory profile of coffee, such as the plant (species, genetic composition, resistance and quality), morphology and anatomy of the fruit, environment (edaphoclimatic conditions), field (crop care, harvest and post-harvest), processing, storage, roasting, grinding and method of preparation of the drink. Thus, it is essential that all stages of coffee processing and preparation are monitored to achieve a beverage with pleasant sensory characteristics.

There are more than a thousand volatiles produced by the roasting phase related to aroma, but only 20 to 30 individually are considered important. It is during roasting that aromas are formed by pyrolysis of water-soluble compounds such as sugars, amino acids and trigonelline. The lipid fraction, on the other hand, undergoes minor changes, forming a protection of the aromatic compounds, since they are retained by the lipid fraction. Some reactions that occur in roasting forming volatile compounds are: Maillard reaction, degradation of trigonelline, degradation of phenolic acids, degradation of lipids, caramelisation of sugars, degradation of sulphurous amino acids and degradation of hydroxy amino acids.

The volatile compounds present in coffee can be divided into heterocyclic and aliphatic.

Heterocyclics are cyclic compounds containing one or more different carbon atoms in the ring. Aliphatics or alicyclics are usually found in lower concentrations compared to heterocyclics. Some examples of hererocyclic (Table 3) and alicyclic (Table 4) volatile compounds are:

Table 3: Heterocyclic volatile compounds in coffee

Compound	Curiosity	Obtaining	Aroma
Furans	Found in large quantities in roasted coffee.	Degradation of carbohydrates (sucrose), occurring by pyrolysis (caramelisation) or by the Maillard reaction. Arabinogalactan is a polysaccharide also considered as a precursor.	Sulphur, old roast coffee, fresh roast coffee, burnt matter, ether, grass, caramel and burnt sugar.
Pyrrhic	Characteristic of thermally processed foods.	Maillard reaction, Strecker degradation, pyrolysis of amino acids and trigonelline degradation.	Sweet and slightly burnt, beef caramel, roasted coffee and old roast coffee.
Oxazoles	One nitrogen atom and one oxygen atom.	Hydroxy amino acid derivatives.	Sweet and nutty.
Thiazoles	One atom of nitrogen and one of sulphur.	Maillard reaction between sulphur amino acids and reducing sugars.	Vegetables, meat, roasted matter and nuts.

Thiophenes	Important for the aroma of coffee, but present in small quantities.	Maillard reaction with sulfur amino acids.	Onion, mustard, sweet, honey, caramel and sulphurous.
Pyrazines	Its precursors are protein-bound amino acids such as threonine.	Maillard reaction, degradation of Strecker and pyrolysis of hydroxy amino acids.	Grass, green coffee, peas, potatoes and raw vegetables, burnt matter, neighbourhood oil, nuts and butter.
Pyridines	Found in foods subjected to heat treatment and microbial activity. Higher amounts in darker roasts.	Maillard reaction, Strecker and trigonelline degradation.	Old roasted coffee, hazelnut, rubber and burnt matter.

Table 4: Aliphatic volatile compounds in coffee

Compound	Curiosity	Obtaining	Flavour
Phenols	Lower concentration, which increases in darker roasts. Higher concentration in	Degradation of free phenolic acids.	Burnt matter, spices, clove, smoke, bitterness and astringency.

	robusta than in arabica.		
Aldehydes	Found in large quantities in fresh coffee. They are lost in storage	Lipid auto-oxidation and reaction of Maillard.	Acres and pungent, light fruity and floral.
Ketones	Decreases during storage by volatilisation. Has a great impact on the final flavour of the drink.	Sucrose pyrolysis and fatty acid auto-oxidation	Fruit, butter, burnt sugar.
Alcohols	Mainly ethanol and methanol.	Degradation of lipids.	Floral, mouldy, honey, beer, sweet, soupy and cat sweat.
Hydrocarbons	No impact on the coffee flavour.	Lipid oxidation and Strecker degradation.	-
Main carboxylic acids: acetic, formic and propanoic.	Many volatilise during roasting. They are not considered to have a high impact on the final flavour.	Degradation of glycans.	Cheese, butter, vinegar, slurry, cream and chocolate.

Esters	Most formed in the fruit before roasting.	-	Pine, apricot, pear.
Compounds sulphurised	Sulphur-containing compounds are of great importance for the aroma of roasted coffee.	Degradation of sulphur amino acids and Maillard reaction.	Putrid.

The aroma that each class of volatiles will present will depend on which compound was formed and its concentration. The aroma profile of coffee is complex as can be seen, being composed of numerous volatiles of different qualities, some pleasant and others not. The concentration, proportion and influence of one flavour over another affects the final result. In general, the robusta species has higher concentrations of total volatiles than the arabica species.

9. Bibliographical References

CLARKE, R. J. Coffee Chemistry, Vol. 1: Elsevier Applied Science, 1985. 304 p.

ESQUIVEL, P.; JIMÉNEZ, V. M. Functional properties of coffee and coffee byproducts. Food Research International, v. 46, p. 488-495, 2012.

ICO, INTERNATIONAL COFFEE ORGANISATION. World coffee production, 2017. Available at: <http://www.ico.org/pt/trade_statisticsp.asp>. Accessed on: 10 Jun. 2017.

MURTHY, P. S.; NAIDU, M. Sustainable management of coffee industry by products and value addition - a review. Resources, Conservation and Recycling, v. 66, p. 4558, 2012.

Sigma-Aldrich. Available at: <https://www.sigmaaldrich.com/>. Accessed on: 03 Mar. 2017.

I want morebooks!

Buy your books fast and straightforward online - at one of world's fastest growing online book stores! Environmentally sound due to Print-on-Demand technologies.

Buy your books online at
www.morebooks.shop

Kaufen Sie Ihre Bücher schnell und unkompliziert online – auf einer der am schnellsten wachsenden Buchhandelsplattformen weltweit! Dank Print-On-Demand umwelt- und ressourcenschonend produzi ert.

Bücher schneller online kaufen
www.morebooks.shop

Printed by Books on Demand GmbH, Norderstedt / Germany